PLEXITY

How Euler's Identity Constructs Reality

A. S. G. McGeorge

Amphikorn Media

ISBN-13: 9781966796046

Cover design by: Amy McGeorge
Library of Congress Control Number:
Printed in the United States of America

To Brittney G-K, who has grown into the backbone of this family, and who boldly went where I was reluctant to go.

And to Filip, who may yet believe.

lescription of what, at the time, we conceptualized as limensions.'

In these dimensions, we determined there were ren, however they were actually linked in a adoxical manner. The dimension zero and 7 were ally a single, conjoined dimension, which together amed the D0/7. The zero represented the seeds ometry. The seven represented the 'degrees of om,' as we used the term, or the infinite potential which those seeds of perfect order must, by ion, arise.

'hen came the D1/6, the intertwined dimensions he 1 represented those points of order now ip into two points connected by a straight line, tion of a one-dimensional state, yet still within set of potential, which was now constrained that order was emerging.

D2/5 now introduced a two-dimensional where the one-dimensional lines now h that each point now had two other points l interact with, and the loop they created new structure, a plane.

/4 then was the three-dimensional state, gle loop now referenced across itself, with e to interact not only with the points djacent, but a third point across the

CONTENTS

INTRODUCTION

I have always found myself fascinated
of things.

The origins of life. The beginnings
The seeds of human language.

One thing I should clarify right
not a theoretical physicist, nor a ma
a genius level Mensa intellect. My
biological sciences, with a smatter
disparate interests. When my
would come, or when I would se
a You-Tube video about physic
jump to that topic first. After
in a fair amount of depth.
sense. My mind preferred n

The roots of this n
one evening during a disc
school aged son. He w
had become intrigued
By the end of our d
living room and to
evening had turne

loop. This defined the state we were familiar with, the geometry of the sphere.

Over the next years, we returned to this topic now and then, refining the 'points of position' and how they might interact with each other. How this geometry might describe the larger universe. How these interconnecting waveforms would interact in creating atomic structure, the properties of a neutron, or of a star, or time.

For myself, I continued to expand on my own hypothesis about the nature of gravity, attempting to integrate it with our dimensional analysis where everything was a waveform interacting across the loop. Gravity not as a field, but as a quantum probability state.

Then came a seemingly overwhelming challenge. How could I represent this model which we had been developing over fifteen plus years, mathematically? I didn't have that skill set. And my son, who does have those skills, had moved on in life. Where could I gain a lifetime of mathematical wisdom, when I couldn't even perform a simple line of calculus?

AI.

I had avoided AI for years. I liked to digitally paint so was disheartened to see how AI was pirating people's lifetimes of experience. I was a writer, an author, and hated thinking that all my work would be subsumed by someone cranking out AI drivel. But when it came to math, I finally succumbed. I began by asking online AI critical questions. However, I wanted my work to be my own, so I set up a personal LLM that could act as my assistant in researching mathematical models.

It was a rocky start. And, just to clarify, while I used AI as a tool and a sounding board, I did not utilize one to write this book. While it assisted in refining ideas, providing better terminology, testing my concepts against real world physics, and informing me about projects in line with my own work, I intended this work to be my own.

Back to the project.

Initially, various online and my own language learning model offered a few mathematical equations as possibilities which we spent hours, and hours, working to develop. Information Geometry. Tsallis non-extensive entropies. Fourier diagrams.

I had no idea what these mathematics were 'telling me.' I couldn't interpret them. I couldn't own them. They were still messy. Still high level. Each time I would ask an AI for a suggestion for a mathematical representation of the burgeoning model, it offered another 'messy' math.

Until one day, I asked for something simpler.

Much, much simpler.

That was the day AI offered up Euler's Identity.

I am not a mathematician; did I mention that? I wasn't familiar with Euler's Identity. I never would have found it alone. When I saw the Identity, sitting there on the screen I didn't see an equation which needed solving. I saw a system, in perfect balance, and how it interacted. At least, how it could interact, the potential inherent in the expressions. I saw elegance, simplicity, yet power. Everything I had been seeking.

I grasped that equation and began to wring every drop of information from it that I could. I asked questions of various AI models and received feedback including hints and descriptions of how the math itself worked, what the expressions revealed. With each discussion, the original D0/7 concept evolved, grew, morphed, until it was an entirely new, much more powerful beast.

I knew not to trust AI. My locally installed home version had made that painfully clear. I was beginning to see its patterns, when it ran out of tokens and lost the plot, or when it just made up something to fill the space. Worse, it proved obsessively sycophantic, gushing about how wonderful I was. Ego stroking was not what I needed. I needed reality checks. Cold water thrown in my face.

Following my trained scientific practices, I began running iteration after iteration through various AI models. I prompted the AI to act like a skeptical theoretical physicist, forcing me to justify every concept. I ran iterations where I didn't explain the model behind the equations, only the equations themselves, allowing the AI to reveal what it could pull from its knowledge base without any leading notes from me. I ran iterations where I did explain my interpretations, the AI's role was to poke every hole.

I bought a decent graphing calculator, knowing how faulty AI could be at solving equations. I tested its calculations at every turn. I read up on the mathematics it mentioned, because I needed to comprehend what I was intending to describe. Throughout all these trials and tribulations, I learned a great deal about mathematical models and theoretical physics and the state of modern research, but I was still far behind the curve.

With each iteration, each discussion, my comprehension changed, expanded, evolved. The original bare-bones D0/7 model we had launched that one evening transformed into something bigger. A more coherent system detailing waveforms and their interconnected system of interaction, built on a powerful mathematical suite of expressions.

The initial rough hypothesis became a comprehensive theory. New levels emerged as the mathematics expanded. All my conceptualizations about the nature of matter, energy, cosmology, gravitation, found a home within its boundaries.

Out of this exploration, three primary axioms framed and directed the process.

First: A subatomic particle, without any other interaction, displays no properties. Or in my favorite descriptive version, 'an atom all alone in the universe has no properties.' Meaning, only during a state of interaction do what we consider 'properties' exist. Only when a 'thing' is in relation to another 'thing' does it 'collapse' into what we experience as reality.

Second: Everything is built from waveforms and the probability of their interactions. The potential for 'existence' underlies everything, while chance determines if that potential is realized.

Third: Euler's Identity is the seed of this hierarch of complexity. The cornerstone upon which all the subsequent levels are built In other words, ahem, 'one equation to rule them all,' if you will.

Euler's equation appears everywhere, in quantum

physics, cosmology. Many research groups are actively investigating the trunk, the legs, the tail. This model of increasing complexity presents a charcoal sketch of the entire elephant.

To coin a term which embodies the intertwined paradox of potential and complexity, a hierarchy of the fundamental scaffolding of 'reality,' the 'tree of emergence,' what follows is:

Plexity.

PLEXITY ONE

The Seed

A seed represents a beginning.

A seed encompasses potential. It is not a tree, yet the pattern for a tree lies dormant within it. And just as a seed is the potential for a new organism, so too, is the seed of Plexity.

I am a biologist by training. I see patterns, emergence, complexity from simplicity, interrelationships, cycles and patterns. This was a natural metaphor.

Let's examine this seed of Plexity.

Euler's Identity equation.

$$e^{i\pi} + 1 = 0$$

Where:

e – the base of the natural logarithm

i – the imaginary unit where $i^2 = -1$

π – the ratio of a circle's circumference to its diameter

1 – the multiplicative identity

0 – the additive identity

At first glance, it might just appear to simply describe the relationship between the most basic, elemental constants. And, so it does.

True enough. Yet, when I first had the equation presented to me, I saw something else.

e – scalability, probability, information

i – imaginary into reality and reality into imaginary

π – waveforms and oscillations, harmonics

1 – stability, completion

0 – cancellation, *no-thing*

I saw a system in perfect balance, represented by the total summing to zero. But that wasn't what intrigued me. I saw a set of perfect potential, and within that set was the result of that potential, a single expression that could blossom into order, a single, perfect point.

That outcome would be expected after all. By definition, any infinite set of absolute potential must include at least one, and potentially an infinite number of, points of absolute, perfect order.

Realize at this level, this initial state, everything is still merely a possibility, not a reality. Yes, even the point of perfect order is purely conceptual, not a reality in object or action. This is the nature of Plexity Level One, or P_1.

Order and potential are not opposites, in this view, rather they are the logical outcome of each other. It's the concept of *no-thing*. In a state of infinite, absolute, pure potential, no-thing can exist. Yet, in any state of perfect potential, by its very definition, there must exist at least one point of absolute, perfect order. Same goes for a state of pure order, where all things become one thing, then again there is no-thing, especially when we consider that properties, things, only exist in relationship to each other. That quantum physics construct about the observer and the observed.

I initially wanted to force each expression into representing one aspect of this infinite zero state, either absolute order, or absolute potential. It's natural to want to codify concepts, slot ideas in this-or-that. This clearer view had to evolve, letting the math lead the way into deeper 'truths.'

Let's look at that mathematically: if we resolve Euler's identity, their sum resolves into zero. No-thing. The system requires both to become meaningful. A zero state has no properties, no structure, no relationship to anything other than itself. Within the expressions lay the seeds for interaction, for stability, for expanding potential. All told, this marked the cornerstone of complexity.

Each of the system expressions expression could represent the seeds of order, the '+1' the total of all possible states, the sum of the set, while $e^{i\pi}$ could represent the seeds of order, the potential turned into reality, the scalability of 'thing.' Viewed another way, the '+1' could represent the seed of order, of stability, while $e^{i\pi}$ represents the seeds of all potential, capable of producing infinite structure.

The point is, this simple, powerful equation describes a state of paradox, where both are true. Pure potential and pure order, all within the same probabilistic set. Like the particle/wave nature of quantum mechanics exists in a state of paradox, here we find those seeds as well.

We can try to break down these latent properties systematically, because our human brains do love the comfort of organizing concepts into little boxes, as follows:

<u>Geometry</u>: seeds of form
<u>Information</u>: seeds of opposites
<u>Trigonometry</u>: seeds of waveforms and harmonics

Quantum: seeds of properties
Relational: paradox of potential
Technological: potential for data
Set: of all potentials including perfect order, the 'infinite zero'

PLEXITY TWO

The Seed Splits, Roots Emerge

I stared at Euler's Identity, knowing my math skills were hardly up to the task, so I asked myself 'what is the simplest operation I could perform on this equation?' Maybe it needs to relate to something, to itself. With a literal shrug, I asked to see Euler's Identity multiplied by itself.

$$(e^{i\pi} + 1 = 0)^2$$

You will note that from this point, I am ignoring the '= 0' sum of the equation. That is by design. In this interpretation, that value simply verifies that the entire system remains in external balance. I intended to investigate what was happening within the system, the relationships arising between the expressions themselves.

I decided to study two primary mathematical expression suites, the binomial expansion expression, and the trigonometric expanded expressions. Technically, there are at least three relevant expansions which could provide direct information regarding the internal system relationships, however one of those which kept showing up was combinatorics. I hadn't heard of combinatorics, or didn't recall if I had encountered it previously. I have AI to thank for its introduction.

At first combinatorics in relation to this suite of equations seemed like hand-waving or sheer numerology. As I explored further, there was validity to the information it brought to the system, especially across plexity levels. It was describing how the expanded coefficients could represent the count or possible configurations. It could be used to derive probabilities. Combinatorics in concert with the trigonometric oscillations appears to be a critical sorting agent during complexity. However, ill equipped as I was to comprehend its potential, I chose to avoid detailed discussions for the purposes of this introduction.

Binomial expansion: At least I recalled this mathematical concept from my years of algebra. I knew information about the system should reside here. Initially I focused my attention primarily at these expressions, especially the relationship of their coefficients.

$$e^{2i\pi} + 2e^{i\pi} + 1$$

Trigonometric: I noted how oscillations and waveforms kept rearing their heads during the expansion process. I wanted to view those expressions as well. Late in the exploration I began to experiment with graphing these waveforms, especially based on their cosine, their 'real' expressions, and sine, or unreal, suites of expressions. Perhaps someone with more skill will recognize patterns that I was unable to discern, especially on my little handheld graphing calculator.

Real (cosine) part:

$$\cos(2\pi) + 2\cos(\pi) + 1$$

Unreal (sine) part:

$$\sin(2\pi) + 2\sin(\pi)$$

What emerges from this level of complexity, P_2? The binomial coefficients tell us the system has broken symmetry, a pivot point around a slightly stronger central value. An imbalance in the perfect symmetry of the infinite zero has been introduced.

The initial state of pure potential now has its first, simplest property emerge. Excluded from the state of pure potential by their very existence, these points aggregate, interact. Duality, pairs of opposites rise from the potential.

Each 'point' shares a relationship with only one other point. Two points connected form a simple structure, the one-dimensional line. For now, these emergent properties could be interpreted as the first hints of polarity, of charge.

When plotting the trigonometric expressions, we see the offset wavefunctions, indicating the waves are slightly out of phase. Movement, the potential for interaction, for interference patterns. The equations reveal the seeds of oscillation, of waveforms.

Together, these are proto-properties, which will only come to fruition at subsequent plexity levels.

<u>Geometry</u>: two points connected by a straight line, one-dimensional
<u>Information</u>: binary, duality
<u>Trigonometry</u>: we can plot the trigonometric expressions and see their wave functions
<u>Quantum</u>: the seeds of dipoles, charge
<u>Relational</u>: each point has only one other partner
<u>Technological</u>: opposing charge as binary
<u>Set</u>: of actualized potential bounded by unactualized potential

PLEXITY THREE

The Shoot Emerges, Roots Deepen

I kept going. After all, that initial operation had provided some intriguing information. What might the next exponential leap reveal?

One important caveat to keep in mind, this system is not describing an evolution over time. In fact, time doesn't yet 'exist' at this level. Nor does mass nor energy. This is a system describing a hierarchy of complexity, how Euler's Identity encodes for increasing complexity and potential, their interactions within the system itself.

Let's examine the mathematical expressions of level P_3.

Binomial:

$$e^{3i\pi} + 3e^{2i\pi} + 3e^{i\pi} + 1$$

Trigonometric:

Real (cosine) part:

$\cos(3\pi) + 3(\cos)2\pi) + 3\cos(\pi) + 1$

Unreal (sine) part:

$\sin(3\pi) + \sin(2\pi) + 3\sin(\pi)$

When we examine the expanded expressions, the binomial coefficients indicate a more balanced internal state, the '1, 3, 3, 1.' The internal '3's' represent a strength, a stability, mirroring each other.

Where in the P_2 each 'point' had a single other point with which it shared a relationship, now each point has 2 relationship, connected to points on either 'side.'

If we dare attempt to map this on to emerging quantum properties, which I do dare, we find here the initial proto-dipole, the '3, 3,' are bounded by the '1,1.' This hints at their ability, or their necessity, to self-assemble. This assemblage would create a single, massive loop, or string. This loop hints at three major concepts:

One: The loop is dynamic, a massive, near infinite string, a set of stability bounded by simpler, more persistent waveforms.

Two: The 'running' of the proto-dipoles

introduces a definite chirality into the system, a handedness that continues to manifest itself throughout the emerging properties.

Three: The itself loop will impact subsequent levels, introducing a persistent property we can express as a bonding, a baseline wave, even the seeds of the strong nuclear force. We'll see more about that at P_4.

Realistically, there could be a single, near infinite loop of these assembled dipoles. Or there could be an infinite number of near infinite loops, or even smaller loops. Or maybe, in a strange attractor fashion, this single loop pulls all other potential loops into itself, such that there exists only a single primary loop. Maybe, like watching a lava lamp, these dipoles are continually merging with and breaking off from the main loop, but the primary loop persists.

Personally, I prefer the latter option. My brain has yet to accept a multiverse view.

That does not make my interpretation the *right* one. After all, my conceptualization of this model has evolved and shifted dramatically over its evolution. But I digress.

The P_3 introduces a two-dimensional structure to the hierarchy. The 'space' enclosed by the loop, no matter how warped or perturbed it might appear, still defines a single plane. Oscillations become waveforms, points become relational origins along the loop.

Organizing our emerging properties for the P_3:

Geometry: the entire loop now defines a two-dimensional space, a plane
Information: each point now has a relationship with two other points
Trigonometry: oscillations emerge
Quantum: the seeds of the strong nuclear force from the bonding of the 'points' of the loop, introduction of chirality
Relational: each 'point' now can interact with two other points
Technological: the 'wiring' on which the digital coding can 'run'
Set: of the intersection of duality

PLEXITY FOUR

The First Leaves

The P_4 is where the real magic starts to emerge. The 'quantum realm,' if you please. Let's examine the expansions themselves, first.

Binomial:

$$e^{4i\pi} + 4e^{3i\pi} + 6e^{2i\pi} + 4e^{i\pi} + 1$$

Trigonometric:

Real (cosine) part:

$$\cos(4\pi) + 4\cos(3\pi) + 6\cos(2\pi) + 4\cos(\pi) + 1$$

Unreal (sine) part:

$$\sin(4\pi) + 4\sin(3\pi) + 6\sin(2\pi) + 4\sin(\pi)$$

Through the lens of combinatorics, we note how the P_3 '1, 1' interact with the central '3, 3,' forming the '1, 4, 6, 4, 1' coefficients of the P_4. This system has balance, but it also has a central strength, a pivot point, in the '6.' This imbalance indicates a sense of 'movement' within the system, not the more static stability we saw in the P_3.

Here, I'll go out another limb. Should we take the coefficients at face value, we have the boundary '1, 1' representing the dual properties of light, the left side '1' the wave like, field like, persistent state, and the right hand '1' the photonic, interactive state. The '4, 4' coefficients represent the baseline suite of leptons, while the '6' represents the six 'flavors' of quarks. The trigonometric expressions support this, especially the final expressions of both the cosine, or 'real,' and the phase shifted sine, or 'unreal,' suite. Thus, the P_4 introduces the most basic properties of the quantum states.

Note that the lepton states mathematically and geometrically can interact with the 'light' states, but the light states cannot interact directly with the quark states. Here we can note the seeds of mass and energy, where the primary origin for the mass of quarks is due to the three-dimensional nature of the 'knot' they create during their waveform interaction, thus their three primary properties, which will be even further expressed as we shall see in P_5 in the formation of the proton.

This 'knot' could be conceptualized of as

'dragging' some of the P_3 strongly bonded loop directly into the structure of the quarks. All told, this could explain the experimentally derived nature of the strong nuclear force, while the seeds of the weak nuclear force arises here from the remaining vibrational state of the locking of the waveforms themselves.

This can illustrate why quarks are 'massive' and leptons are 'light.' Leptons are not a tripartite knot, rather encoding a dualistic nature. Leptons are not tightly bound within a 'knot,' thus mathematically and structurally available to interact with electromagnetism.

Here the mathematics also indicates a baseline structure for the quantum fluctuations of the foam. Bounded by the properties of light, the system is now based on the harmonic probabilities of interaction. The combinatorial richness of the across the P_2, P_3, through the P_4 levels indicates the potential behind the 'vacuum.' This potential drives quantum particle creation where the waveforms of the P_3 interact self-referentially across the loop. This creates the quasi-particle/waveforms, which as we will see, only resolve, or collapse into states with defined, stable properties, when able to interact with another quasi-particle/waveform.

While the quantum foam contains the potential

for vast energies and masses, the actual probability of two waveforms 'bubbling' up within a particular distance, possibly the Planck length or similar, where they can interact and thus stabilize each other into properties, is very, very low. The probability of forming a tripartite structure is even lower, evidenced by the actual ratios of light, leptons, and bosons in the universe.

Thus, this also addresses the magnitudes of difference between the theoretical cosmological constant, and the measured cosmological constant. One contains all the potential energies combined, while the other, the reality, is due to the extremely low probabilities of emergent stabilization.

If mass of quarks is represented by the cosine expression, the 'real' portion of the trigonometric expression, then the waveform, motion-like properties are the phase shifted 'imaginary' sine expression, which could well represent the relationship between mass and energy, these seeds also emerge from the P_4.

Each point of the loop now can share a relationship not only with the points to either side, but also across the loop. Which 'points' interact is a dance of probability. Some 'near' regions attempting to interact might create strain on the loop itself, impacting subsequent interactions, perhaps altering the 'laws of physics,' or leading to unusual, localized events.

What we see emerging from the P_4, then, is the origins of three-dimensional space. Until this point, such a geometric configuration was merely potential. Here, the near-infinite loop of dipoles becomes self-referential, such that the waveforms created by the harmonic properties. This constructs the basic space with we are familiar in the form of a 'sphere.'

We will also find quantum time and quantum gravity here. Their full states, as with all these seeds, will not be actualized in the forms with which we are familiar until higher plexity levels, here we find their origins.

At this point, I'll go even further out on a limb and posit that gravity is a probability between two (or more, many, many more) waveforms that their harmonics would cancel.

Mind you, this is not an actual cancellation event. It is a statistical probability that the waveform structure would 'cancel.' This should then create a pseudo-Bose-Einstein-condensate moment, a state where the two waveforms attempt to occupy the same state. In a complex body of any real size, they obviously cannot physically occupy the same wave function nor the same space, rather the attempt collapses the three-dimensional interaction towards a more one-dimensional space.

Between two isolated waveforms, such as two electrons, this probability is extremely low. Increase the number of waveforms available for this cancellation event, and the probability becomes an actuality. The process may even stabilize across the body of matter, becoming self-perpetuating. I conceptualize the overall process as analogous to van der Waals in chemistry, a weak local force, however once a critical mass is reached, the effect becomes a self-sustaining probability that could even 'flow' throughout the gravitational body. The more 'massive' the body, the more waveforms available to statistically cancel, the greater the pull towards that center point.

An interesting result of this investigation leads to an intriguing conclusion: the opposite 'pole' of quantum gravity must be quantum time.

In quantum time, time 'ticks' forward only during a waveform changing state due to interaction with another waveform, such as an emission, an absorption, a change of energy configuration, a bond structure shift. If a photon experiences no time between emission and absorption, and if light is the boundary state, then only during novel interactions does time exist.

How does this make gravity the opposing pole to

time?

Quantized time requires waveforms available to participate in novel interactions. When vast numbers of waveforms are participating in statistical cancellation, they are not available for fresh interactions. Thus, time increasingly slows closer to a gravitational body center point.

Let's summarize some of the properties emerging at the P_4.

Geometry: three-dimensional space is created by the spherical potential of the loop
Information: this is the state where the waveforms can begin to encode information, a baseline carrier wave
Trigonometry: full return of waveforms, complete oscillations
Quantum: properties of quantum mechanics
Technological: the operating system, directs the base machine structure
Relational: each 'point' can now interact with three other points, the two to either side, and one across the loop
Set: of all combinations

PLEXITY FIVE

Reaching for the Light

Here, we note how the expanded expressions of the P_4 can interact to create the properties of the P_5. Once again, the system is in a state of equilibrium, a balance built on mirroring of properties. Thus, we can see the emergence of the conservation laws, the 'equal and opposites.'

Binomial:

$$e^{5i\pi} + 5e^{4i\pi} + 10e^{3i\pi} + 10e^{2i\pi} + 5e^{i\pi} + 1$$

Trigonometric:

Real (cosine) part:

$$\cos(5\pi) + 5\cos(4\pi) + 10\cos(3\pi) + 10\cos(2\pi) + 5\cos(\pi) + 1$$

Unreal (sine) part:

$$\sin(5\pi) + 5\sin(4\pi) + 10\sin(3\pi) + 10\sin(2\pi) + 5\sin(\pi)$$

The P_5 is where the interactions between the

properties established in the P_4 interact to create the 'rules of the game.' Opposing forces emerge. The most basic conservation laws, conservation of charge, linear momentum, angular moment, energy and mass. Nuclear physics, the creation of nuclei. Perhaps, as hinted from the discussion of the P_4, even gravity and time, just now represented closer to the field or flow state as we experience them.

The mathematics hints how nuclear chemistry emerges as the quarks from the '6' of the P_4, their combinations forming the properties of protons, mirrored by the '10, 10' from the P_5. The P_4 quantum state behaves like a carrier wave, expressed in the P_5 where more dense information can be encoded. The binomial expressions and trigonometric expressions reveal how light still retains the potential to interact directly with lepton, while its properties remain as the system boundary.

The 'points' of the P_3 loop are not only interacting across the loop to define their P_4 quantum properties, they are also changing their interaction states. We often consider this the dimension of 'time.' The loop itself can now behave as a carrier wave, its interactions acting as algorithms, compressing information. These interactions further refine the properties, the potential for interaction increasingly entwined with the act of interaction.

Pascal's triangle emerged, but that is to be expected from any binomial expansion. Nonetheless, might that pattern prove to have greater meaning derived from this model?

I pause here for a speculation arising from the model so far which could relate to cold fusion: it is possible that the tunneling which allows protons to join into a nucleus, despite the repelling force of their identical charge state, is modulated not by high energy, rather by a similar probability to gravitation, such that when the quantum wave states statistically cancel, this allows the protons to merge during that 'moment' of cancellation of waveforms. Due to the 'knot-like' nature of quarks, this cancellation probability would be far lower than that between waveforms such as electrons, but never zero.

<u>Geometry</u>: the seeds of time emerge in the P_4 realized in the geometry of time, the flow we experience from statistical accumulation of quantum ticks
<u>Information</u>: more complex coding along the baseline carrier wave, compression algorithms
<u>Trigonometry</u>: the system displays opposing waveforms, balance
<u>Quantum</u>: waveforms interact in stable sets of

determined properties
Relational: each 'point' now interacts dynamically with 3-plus-1 points during the change of state
Technological: the application layer, where the 'rules of the game' are set, the layer which codes how the P_4 is allowed to interact, the game engine, the router
Set: of dynamic sets

PLEXITY SIX

The Trunk Strengthens, the Roots Run Deep

As expected, here the expressions of the P_5 combine to create the properties of the P_6. Mathematically, the system is once again 'in motion.' Waveforms are interacting at macro levels. Statistical interactions reign. Stability and motion. Energy creates change. Atoms assemble into complete structures with their valence energy and chemical bonding properties. The twinned aspects of inertia. Thermodynamics, fields and tensors. Fluid Dynamics. Time and gravitation, energy and mass, all interweaving in the creation of stars and planets, defining their complex orbits. Plate tectonics, erosion and uplift.

Again, we are visited by paradox. It seems odd that the P_6 would encode both stable atoms as well as galaxies, however the increased range and value of binomial coefficients express this, providing a wide range of large-scale interaction. Light persists, the '1, 1' bounding coefficients remain, the final trigonometric expressions describing its new potential for interaction.

Binomial:

$$e^{6i\pi} + 6e^{5i\pi} + 15e^{4i\pi} + 20e^{3i\pi} + 15e^{2i\pi} + 6e^{i\pi} + 1$$

Trigonometric:

Real (cosine) part:

$$\cos(6\pi) + 6\cos(5\pi) + 15\cos(4\pi) + 20\cos(3\pi) + 15\cos(2\pi) + 6\cos(\pi) + 1$$

Unreal (sine) part:

$$\sin(6\pi) + 6\sin(5\pi) + 15\sin(4\pi) + 20\sin(3\pi) + 15\sin(2\pi) + 6\sin(\pi)$$

In this model, P_6 is also the realm of all our 'messy' mathematics, our calculus and linear algebra, matrices, Hamiltonian operators, General and Special Relativity. All these are our attempts to complete the picture of that elephant, poking at the concepts using systems evolved over time, as our own conceptualization have evolved. From this model's view, all these various fields physics and cosmology are attempting to peer through the P_5 layer in order to explain the P_4 and below.

Here, my comprehension of mathematics fails me. There is some filter that a fifth level binomial expansion creates that limits the information between the lower exponents and the higher. This property seems relevant to the overall system, to our mathematics itself, but I am forced to admit that I didn't understand that property well enough to defend it, though I think it is relevant to our 'messy' mathematical approximations of the lower plexity levels.

At this plexity, the probabilities defining gravitation and flow of time are in full flower. I have made multiple attempts to define a probability that would map directly onto measured values of gravitation. I attempted to enter that realm of 'messy math,' and failed.

One of the greatest challenges I faced, obviously, is that this model is based entirely on waveforms. There is no measure of length, or time, mass, energy or charge, as we experience them at the macroscopic level. These are all emergent properties. Therefore, translating this directly into measured units was beyond my skill. I must leave it to greater minds than mine to develop such conversion factors.

Still, perhaps we might we derive thermodynamics directly from these equations? Can we assign properties directly to every expression of every complexity level? Can the three-dimensional graph of the P_4 trigonometric expressions give direct evidence for the probability curves associated with property stabilization?

In that light, I attempted to derive some of the foundational unitless constants from first principles of this suite of Euler's Identity equations. These results are explained more fully in the chapter by that name. Only then will we see just how this model really stacks up against full on reality. So far, it has been holding its own quite well.

Upon further reflection, I began to wonder if the binomial expansions weren't revealing structural information regarding the 'particle' nature of the 'stuff' that makes up existence. Which leaves the trigonometric expressions describing the 'wave' nature. Perhaps hints about the relationship between matter and energy. Dare I hope this model could provide first principles equations for those as well?

Summarizing the analogies sorted by domain:

<u>Geometry</u>: fractals, crystals, chemical bonds, electrons energies

Information: error correction and stable decoding, complex mathematics
Trigonometry: increased complexity of waveforms, waveforms within waveforms
Quantum: macroscopic behavior from the total of quantum states, macro gravity, time dilation
Relational: interaction becomes statistically driven, 3-plus-1-plus-many
Technological: the user interface, the web browser or game itself
Set: of stable patterns, macroscopic properties

PLEXITY SEVEN

The Tree Flowers

P_7 is where I posit the emergence of life.

Life, the complex intersection of chemistry, thermodynamics, time, gravitation, light, mass and energy. Life at its most reductive utilizes the tools of chemistry to harness physics. Maintaining its internal complexity by increasing its local entropy. Little anti-entropy machines engaging with the outside world, reorganizing their external environment in new ways.

The seeds of internal complexity bloom here. New, even more complex emergent properties abound. Information had a form, now it takes on function. Potential for fractal like interactions. Feedback loops. Clock reactions and cycles. Membrane potentials and transport proteins. Cells within cells. Circadian rhythms. Metabolic networks as well as entire ecosystems engaged in an intricate dance around cooperation and competition for resources.

Binomial expansion:

$e^{7i\pi} + 7e^{6i\pi} + 21e^{5i\pi} + 35e^{4i\pi} + 35e^{3i\pi} + 21e^{2i\pi} + 7e^{i\pi} + 1$

Trigonometric:

Real (cosine) part:

$$\cos(7\pi) + 7\cos(6\pi) + 21\cos(5\pi) + 35\cos(4\pi) + 35\cos(3\pi) + 21\cos(2\pi) + 7\cos(\pi) + 1$$

Unreal (sin) part:

$$\sin(7\pi) + 7\sin(6\pi) + 21\sin(5\pi) + 35\sin(4\pi) + 35\sin(3\pi) + 21\sin(2\pi) + 7\sin(\pi)$$

The central '35' coefficients offer a large array of states, an explosion of potential interactions. The distribution of coefficients is broader, gentler. The trigonometric waveforms are even more complex, with new possibilities for emergent properties. Together, these equations describe a system in complex, dynamic balance.

One of the analogies I have followed through this discussion related to information technology. The lowest plexity levels introduced the machine language, the basic hardware. At P_4 the function resembles an operating system, the program which manipulates the hardware levels into storing and analyzing binary data.

The P_5, then, resembles the gaming engine, the 'rules of the game,' the application sitting just above the operating system. The P_6 is then the user interface, the gui, the visual game. Which leaves us here, the P_7, to represent us, the user.

I had hoped to have much more to say about this plexity level. After all, it should be the level describing my own realm of study. Unfortunately for my brain, there's just so much to unpack by the time we reach the P_7. Details of the P_4, P_5, and P_6 still wait to be unraveled. What each expression and their relationships between and across levels reveals. What this model has presented is just an overall sketch of that wonder prompting elephant, not yet a painting, not a sculpture, and certainly not yet a living, breathing being.

While we're investigating this living level, I chose to fall into another rabbit hole: what about extraterrestrial life? This model would lead one to accept that life must be a natural consequence, an emergent property rising out of complexity. And for that, I do think that the chemistry which describes the most basic life is a certainty. However, as we have noted, probability plays a huge factor in every stage, more so for the more complex condition we like to think of as 'higher' forms of life.

For one thing, I think the conditions leading

to 'higher' life are more constrained than we might imagine. They not only require a liquid in which complex chemical reactions can come in contact, mix, and become stabilized, which in its early stages this 'washing' of that liquid across the surface and induce the necessary mixing, as well as challenges to the budding proto-life forcing it to adapt was assisted by our singular large planetary sister, the moon. Such a substantial sister planet provides a secondary barrier from the worst of meteor impacts. Moreover, not only does a large sister planet provide stability to the axial tilt, this smaller companion also as a gravitational counterpart, which could prove critical for maintaining an active planetary sphere, plate tectonics, material exchange, over longer time periods than a system lacking such a cyclic gravitational tug-o-war.

Importantly for our own evolution, such a moon becomes as a beacon, a concrete object for an emerging intelligence to marvel over, something that feels close enough to touch, and dream of reaching. A way station to the stars.

Then there's the fact that the vast majority of exoplanets found to date are gas giants, well within the orbits of our own rocky planets. Is that the norm? Is our star system really so unique after all? If not, why haven't we heard from 'anyone else out there?'

What if the mathematics and symbols we have been popping off into the universe are nothing more than *Homo habilis* banging a stick on a rock? What if this model of complexity, should it prove able to predict totally new ways of interacting with the 'stuff of the universe,' is the least level of communication to which 'they' might bother to respond?

Just a thought.

Geometry: membranes, compartmentalization, dynamic structures

Information: semantics, interpretation, self-modifying, predictive, mutative

Trigonometry: complex feedback loops, signaling, self-regulation

Quantum: observer and observed

Relational: complex suites of waveforms patterning with other suites of waveforms

Technological: the user level

Set: of the biosphere

HIGHER PLEXITIES

Fruits of the Labors

I chose not to go into any detail regarding these levels of complexity. As I've noted, besides the equations themselves, I find there is still such a wealth of information yet to unpack about the P1 through P_7 already. How to define the exact relationship between the expressions of the P4 upon the P5, and the P_5 upon the P_6, for example. Translating the P_6 into real-world measurements. Gravitation as probability.

It is possible the P_8 encodes culture, although I would argue that culture is less unique than humans care to admit, such that everything about living organisms could be encoded by the P_7. Ecosystem, biomes, technology. That is merely my current hypothesis, which has yet to be modified after examination of this level.

Maybe P_8 encodes information and cognition at a planetary level. Or perhaps at a star system level. We could delve deeper into each of the expressions, how combinatorics reveals how the expressions of a level interact to create the expressions of the subsequent level. That would require higher computer modelling power than I have access to!

That leaves P_9, P_{10}, P_{11}? Just how high can this complexity go? Theoretically, it could continue through infinity. However, I posit that at some point, complexity would be so, well, complex, the potential and realized intertwined in such diverse and distinct ways, that it could become indistinguishable from the initial P_1 state. Perhaps at some level the plexity even collapses back into the P_1, forming an infinite looping.

Personally, I like this final plexity is the P_{10}, but that could be a limitation of my own brain. Or prejudices due to cultural familiarity. These levels could exist far beyond this current model, maybe the P_{13}, or higher powers still.

DERIVING UNITLESS CONSTANTS FROM FIRST PRINCIPLES

Yes, I am going out on a limb here, but if this model holds true, it needs to do more than just provide a conceptualization of the hierarchy of complexity. It needs to be predictive. It needs to be mathematically sound. It needs to be powerful enough to derive real world information that could be used function as well as illustrate conceptually.

Of course, I hope that the model proves to be just that useful, that powerful. I hope it can be expanded upon to derive just such explanations of the underpinnings of our reality. That was the purpose of compiling all these thoughts, of pursuing the mathematical equations.

Here are a few examples that might whet the appetite.

Proton-Electron Mass Ratio

Let's investigate the simplest mathematical relationship which I have discovered so far, the proton-electron mass ratio. This describes the ratio between a basic proton, and a basic electron. Recall, the proton is made of three P_4 quarks, each of which draws its mass from the P3 loop, while an electron draws barely any of the properties of the P_3 loop, and is more associated with the properties of the P_2 seeds of charge.

One of the recognized closest approximations is the $6\pi^5$.

$6\pi^5 = 1836.1181$

Real value: 1836.1526

This equation itself can be derived directly from first principles of the P_4, recognizing that other plexity levels are often the 'filter' through which we interact with a lower level.

The '6' represents the quarks, the seeds of the proton. An electron's value would be a 1, which we wouldn't notice, unless that also accounts for the slight variation in the theoretical versus measured ratio. π^5 represents the P_5 level oscillations, the complex waveforms actually making up the relationship between electrons and protons.

This creates an error of, which could be explained by the fact we are still dealing with P_4 and P_5 reality through the P_6 level.

Weak Mixing Angle

This is a more complex relationship. Most closely related to the concepts of this mode, the weak mixing angle describes the three-dimensional angles created in the 'knot' of three quarks forming a proton. It is chiral due to the chirality of the loop which formed the primary mass of the quarks.

In this model, then, it is a relationship between the P_2 charge property, the P_4 three-dimensional knot, and the inverse of the scalability, perhaps indicating how the knot itself folds in on itself.

$$(2/4)^2 - 1/e^4 = 0.231685$$

Real value (at low energies) – 0.22305
Real value (at higher energies) – 0.23153

The fact that this 'constant' can fluctuate is also indicated by the inverse scalability, as well the fact that our mathematics is describing these values through the lens of thermodynamics from the P_6.

Fine-Structure Constant

This last unitless constant I was able to find a correlation, but the derived equation is not nearly so clean. Then again, the definition itself contains a lot more moving parts.

This constant represents the relationship between light and charged electron states. It is a probability whether an electron will emit, or absorb, a photon. It could be thought of as the tension between matter and light, which both emerge at the P_4, which are heavily modulated by the expressions arising from the P_5 and P_6.

We can approximate the value by:

$$(16 \times \pi \times e) + 1/\pi + 1/\pi^2 = 137.0553378$$

The actual value of $1/\alpha = 137.035999$

In this example, the 16 represents all the possible states of the P_4 combined, times the waveform oscillation and scalability factors. Then we add the inverse of π and the inverse of π^2. Here, I stumble. Maybe this is relevant, maybe it is just playing numerology.

Further Patterns Worth Exploring

If these patterns hold true, then other unitless constants should be able to be derived using these first principles. Understanding which levels are interacting, which properties are the seed or the actualization, could provide even more understanding regarding how these properties arise, and how they create reality as we experience it.

I purchased a graphing calculator with the explicit purpose of double checking any and all the equations resolved by an AI, aware that the tool notorious for its lack of formal math skills. It is capable of providing descriptions, definitions, so long as some other human did it first, and AI had merely to parrot back what it scraped. I didn't dare trust it, thus the multiple iterations prompting for expanded equations, rearrangements, just to find the pattern that indicated it wasn't the product of a sycophantic hallucination.

The calculator provided me the ability to dabble in graphing the trigonometric expressions. And I do mean dabble, because I was relying on AI to give me the best expressions to plug in. Sometimes it gave me useful data points.

I graphed total expressions of the 'real' and the imaginary expressions combined. This formed something like a Pacman, facing to the left and its mouth closed. Each plexity level had the same overall form, just the mouth got a little more 'loopy.' A description of the overall system, which scarcely altered over the P levels.

I tried graphing the trigonometric expressions each level, allowing me to compare the various expressions with each other. Intriguing patterns emerged, tubelike structures, phase shifts, nested waveforms. Differing frequencies and amplitudes.

But the real problem lies in my inability to comprehend what the graphs might be revealing. Using a P_4 example:

X(T): T
Y(T): cos(4*π*T)
Z(T): sin(4*π*T)

When I rotated the three-dimensional plot, I see what looks just like a spirograph drawing. Unfortunately, my graphing calculator provided very little detail. Still, I wondered if I might be seeing indications of the probability field directly from first principles. Mostly I felt like Jack Skellington marveling over Christmas town: 'What does it mean, what does it mean?'

Are there any further direct correlations between

the coefficients and measured properties? Their pattern hints at something more to be revealed. Surely someone more versed in theoretical physics and mathematics can visualize other hidden structure within and across the suite of equations.

Revisiting combinatorics, perhaps probabilities can be derived from first principles. That would be as exciting as the potential for probability fields determined from the three-dimensional graphs of the trigonometric expressions.

I must leave the future to other great minds to flesh out this sketch of an elephant, paint it, sculpt it, computer model it.

My greatest hope is that other researchers will find value in this Plexity model. That they will manage to tease out every ounce of potential, build upon it, develop a powerful and solid architecture from these most humble, elegant blocks. Uncover how a fully developed model could describe the underpinnings, the structure behind the structure, for their own works.

May the games begin.

FINAL MUSINGS

Besides just quantum gravity and time emerge during contemplations of this model, so too have other more fringe ideas developed. This was as good a place to posit them as any. I'm also restating some of my previous more speculative ideas emerging from the plexity model.

I have pondered the nature of the P_2 linkage between the emergent duality. In order for these identical points to be in opposition, all it requires is for an identical spin to be 'flipped' over, so to speak. Then, is this bond static, or dynamic? Is it just a simple line 'drawn' between opposing proto-charge states, or does it behave in a more lemniscate, or even a mobius, fashion, such that this emerging charge state flows between the points?

I've wondered about the true nature of the neutron, and its instability when not within a nucleus. What if the neutron is stabilized by a dynamic interaction between the orbital electrons and the 'electron' property, such that they physically were interchangeable, or the waveform 'moved through the neutron?' This predicts that different bonded states of isotopes should display slightly different half-lives, which they do.

What if the quantum foam, with light being an emanation of, interacts weakly with light such that it causes much of what we see as redshifted? Not tired light, because it wouldn't lose coherence, rather 'stretched' due to interaction with the quasi states of waveforms which attempt to form collapsed-into-reality matter, but without a partner waveform those properties never stabilize. Yet light, as a realized perturbation of the quantum state, the simplest most prevalent form of information transfer across the system, should be ever so slightly impacted by the various quasi states of that same baseline.

Which leads to dark matter/dark energy/dark magnetism being properties of the quantum foam activity. This potential is only partially realized due to such low probabilities. Yet, across the vastness of space, these quasi states and the properties which do manage to stabilize, no matter how unlike those probabilities may actually be, might such macroscopic properties of the quantum foam also impact planetary orbital dynamics?

What if the rotation of galaxies was stabilized not just by dark matter, but by time dilation, such that time moves faster farther from the massive black holes at the core, allowing the outer arms to 'keep pace' with the rotation of the core?

While the quantum foam contains the potential for vast energies and masses, the actual probability of two waveforms 'bubbling' up within a particular distance, possibly the Planck length or similar, where they can interact and thus stabilize each other into properties, is very, very low. The probability of forming a tripartite structure is even lower, evidenced by the actual ratios of light, leptons, and bosons in the universe.

Thus, this also addresses the magnitudes of difference between the theoretical cosmological constant, and the measured cosmological constant. One contains all the energies of the quantum states across the entire potentials combined, while the other, the reality, is due to the extremely low probabilities of emergent stabilization.

A little more musing regarding the gravity and time as opposing 'poles.'

I discussed how quantum time, time 'ticks' forward only during a waveform changing state due to interaction with another waveform, such as an emission, an absorption, a change of energy configuration, a bond structure shift. If a photon experiences no time between emission and absorption, and if light is the boundary state, then only during novel interactions does time exist.

Quantized time requires waveforms available to participate in novel interactions. When vast numbers of waveforms are participating in statistical cancellation, they are not available for fresh interactions. Thus, time increasingly slows closer to a gravitational body center point, and time dilation as light attempts to follow a straight line, but the quantum space it is traversing is collapsed, or bent.

You might note here this idea leads to the conclusion that is may not be the nucleus mass itself which is primarily responsible for gravitational attraction. Those nucleons are pretty much locked up, their waveforms intrinsically 'busy.' That leaves the less constrained leptons to participate in both gravitational cancellation events, and time ticking forward.

In regards to that probability of the pseudo-Bose-Einstein-condensate-moment, the PBECM for short, I have postulated the form that probability curve might take. It would start out nearly at zero, and move slightly at low waveform density, increase sharply at some statistical threshold, start to level off sharply at the waveform density of both planets and stars, then rise only slightly as it reaches waveform densities of neutron stars, where it would again rise sharply until it stop just before reaching a probability of 1, this range indicating black hole level gravitation, where the waveforms are in a nearly identical singe phase and harmonic.

How about the 'bubbling' of the quantum foam near a gravitational body? I posit that this activity is not actually inhibited, although when quantum states are already 'locked' in the three-dimensional 'knots' of quarks and nucleons, those waveforms are already in fully realized states. However, it is possible that freer states such as leptons may be actually in a state of dynamic replacement, such that a single electron is not persistent as such, may, even as light is being 'recreated' at each node, falling out of collapsed reality and being recreated at any given quantum moment.

Returning briefly to the question regarding is our own solar system unique, I ponder why our system, as opposed to the vast majority of found exoplanets, does not have a single gas giant within the 'goldilocks' zone. I understand that current theory holds to these planetary bodies migrating inwards toward their central gravitational body, their star, and that our rocky bodies have no gas giant outer layer due to the initial nuclear fusion shock wave blowing away the original atmosphere. I have also noted, however, that when you examine the images of nebulae, you can actually visualize the galactic magnetic force lines where the charged plasma has been pulled into long streamers.

What if the initial nuclear fusion of a star doesn't happen in a massive explosion, rather it bubbles up like boiling water, therefore that process doesn't naturally lead to a shock wave blowing away the larger atmosphere? What if, due to the fluid dynamic behavior of and conservation of angular momentum of the blown off stellar material create a 'blobbing' effect, with two or three or many inner swirls of more dense material around which the entire proto-stellar system alternately revolves, leading directly to the creation of large gas giants near the most prominent star? What if instead, for our own solar system, during a key moment of its stabilization, a line of the galactic magnetic field managed to strip away enough of the lighter, gaseous elements that only the heavier elements remained to form the planetary bodies?

There could well be more patterns, more real-world correlations, across these plexity levels that I have presented here. Other expansions relating Euler's Identity which I have yet to run across. Calculations which might derive directly from the equation suite that I will never see. What might this model mean for physics, for cosmology, for biology, in future studies? If it holds true, what new directions in technology might this lead? Buckminster fullerene matrices designed to create localized gravitation? Manipulation of the P_3 and P_4 properties directly? An entirely new way to communicate through the universe?

Just thought I'd leave with those parting thoughts.

SUITE OF PLEXITY EQUATIONS:

Binomial

Plexity Level	Binomial-Expanded Expression
1	$e^{i\pi} + 1$
2	$e^{2i\pi} + 2e^{i\pi} + 1$
3	$e^{3i\pi} + 3e^{2i\pi} + 3e^{i\pi} + 1$
4	$e^{4i\pi} + 4e^{3i\pi} + 6e^{2i\pi} + 4e^{i\pi} + 1$
5	$e^{5i\pi} + 5e^{4i\pi} + 10e^{3i\pi} + 10e^{2i\pi} + 5e^{i\pi} + 1$
6	$e^{6i\pi} + 6e^{5i\pi} + 15e^{4i\pi} + 20e^{3i\pi} + 15e^{2i\pi} + 6e^{i\pi} + 1$
7	$e^{7i\pi} + 7e^{6i\pi} + 21e^{5i\pi} + 35e^{4i\pi} + 35e^{3i\pi} + 21e^{2i\pi} + 7e^{i\pi} + 1$

Trigonometric

Plexity Level	Real Component
1	$\cos(\pi) + 1$
2	$\cos(2\pi) + 2\cos(\pi) + 1$
3	$\cos(3\pi) + 3\cos(2\pi) + 3\cos(\pi) + 1$
4	$\cos(4\pi) + 4\cos(3\pi) + 6\cos(2\pi) + 4\cos(\pi) + 1$
5	$\cos(5\pi) + 5\cos(4\pi) + 10\cos(3\pi) + 10\cos(2\pi) + 5\cos(\pi) + 1$
6	$\cos(6\pi) + 6\cos(5\pi) + 15\cos(4\pi) + 20\cos(3\pi) + 15\cos(2\pi) + 6\cos(\pi) + 1$
7	$\cos(7\pi) + 7\cos(6\pi) + 21\cos(5\pi) + 35\cos(4\pi) + 35\cos(3\pi) + 21\cos(2\pi) + 7\cos(\pi) + 1$

Plexity Level	Unreal Component
1	$\sin(\pi)$
2	$\sin(2\pi) + 2\sin(\pi)$
3	$\sin(3\pi) + 3\sin(2\pi) + 3\sin(\pi)$
4	$\sin(4\pi) + 4\sin(3\pi) + 6\sin(2\pi) + 4\sin(\pi)$
5	$\sin(5\pi) + 5\sin(4\pi) + 10\sin(3\pi) + 10\sin(2\pi) + 5\sin(\pi)$
6	$\sin(6\pi) + 6\sin(5\pi) + 15\sin(4\pi) + 20\sin(3\pi) + 15\sin(2\pi) + 6\sin(\pi)$
7	$\sin(7\pi) + 7\sin(6\pi) + 21\sin(5\pi) + 35\sin(4\pi) + 35\sin(3\pi) + 21\sin(2\pi) + 7\sin(\pi)$

Plexity	Full complex expression
1	$[\cos(\pi) + 1] + i[\sin(\pi)]$
2	$[\cos(2\pi) + 2\cos(\pi) + 1] + i[\sin(2\pi) + 2\sin(\pi)]$
3	$[\cos(3\pi) + 3\cos(2\pi) +3\cos(\pi) + 1] + i[\sin(3\pi) + 3\sin(2\pi) +3\sin(\pi)]$
4	$[\cos(4\pi) + 4\cos(3\pi) +6\cos(2\pi) + 4\cos(\pi) +1] + i[\sin(4\pi) + 4\sin(3\pi) +6\sin(2\pi) + 4\sin(\pi)]$
5	$[\cos(5\pi) + 5\cos(4\pi) +10\cos(3\pi) + 10\cos(2\pi) +5\cos(\pi) + 1] + i[\sin(5\pi) + 5\sin(4\pi) +10\sin(3\pi) + 10\sin(2\pi) +5\sin(\pi)]$
6	$[\cos(6\pi) + 6\cos(5\pi) +15\cos(4\pi) + 20\cos(3\pi) +15\cos(2\pi) + 6\cos(\pi) +1] + i[\sin(6\pi) + 6\sin(5\pi) +15\sin(4\pi) + 20\sin(3\pi) +15\sin(2\pi) + 6\sin(\pi)]$
7	$[\cos(7\pi) + 7\cos(6\pi) +21\cos(5\pi) + 35\cos(4\pi) +35\cos(3\pi) + 21\cos(2\pi) +7\cos(\pi) + 1] + i[\sin(7\pi) + 7\sin(6\pi) +21\sin(5\pi) + 35\sin(4\pi) +35\sin(3\pi) + 21\sin(2\pi) +7\sin(\pi)]$

PLEXITY LEVEL PROPERTIES

Plexity Level	Geometric Description
1	Seeds of form; primordial potential for structure.
2	Two points connected by a straight line, forming one dimension.
3	A closed loop defining a two-dimensional plane.
4	Spherical potential of the loop giving rise to three-dimensional space.
5	Seeds of time emerging from accumulated quantum ticks.
6	Fractals, crystals, chemical bonds, electron energy structures.
7	Membranes, compartmentalization, dynamic structural organization.

Plexity Level	Information Description
1	seeds of opposites
2	binary, duality
3	each point now has a relationship with two other points
4	this is the state where the waveforms can begin to encode information, a baseline carrier wave
5	more complex coding along the baseline carrier wave, compression algorithms
6	error correction and stable decoding, complex mathematics
7	semantics, interpretation, self-modifying, predictive, mutative

Plexity Level	Trigonometry Description
1	seeds of waveforms and harmonics
2	we can plot the trigonometric expressions and see their wave functions
3	oscillations emerge
4	full return of waveforms, complete oscillations
5	the system displays opposing waveforms, balance
6	increased complexity of waveforms, waveforms within waveforms
7	complex feedback loops, signaling, self-regulation

Plexity Level	Quantum Description
1	seeds of properties
2	the seeds of dipoles, charge
3	the seeds of the strong nuclear force from the bonding of the 'points' of the loop, introduction of chirality
4	properties of quantum mechanics
5	waveforms interact in stable sets of determined properties
6	macroscopic behavior from the total of quantum states, macro gravity, time dilation
7	observer and observed

Plexity Level	Relational Description
1	paradox of potential
2	each 'point' has only one other partner
3	each 'point' now can interact with two other points
4	each 'point' can now interact with three other points, the two to either side, and one across the loop
5	each 'point' now interacts dynamically with 3-plus-1 points during the change of state
6	interaction becomes statistically driven, 3-plus-1-plus-many
7	complex suites of waveforms patterning with other suites of waveforms

Plexity Level	Technological Description
1	potential for data
2	opposing charge as binary
3	the 'wiring' on which the digital coding can 'run'
4	the operating system, directs the base machine structure
5	the application layer, where the 'rules of the game' are set, the layer which codes how the P4 is allowed to interact, the game engine
6	the user interface, the web browser or game itself
7	the user level

Plexity Level	Set Description
1	of all potentials including perfect order, the 'infinite zero'
2	of actualized potential bounded by unactualized potential
3	of the intersection of duality
4	of all combinations
5	of dynamic sets
6	of stable patterns, macroscopic properties
7	of the biosphere

AFTERWORD

I relate the following not because my biography is particularly special, only that it is relevant to how my world view evolved. It is relevant to realize that anyone might develop a model like Plexity, if they had the drive to do so. As I noted, I am interested in the root of things, the beginnings, the 'whys.' Therefore, I thought a trip through my own roots might be fitting.

Tolkien once wrote of Smeagol that '*He was interested in roots and beginnings.*' He stated this as though that was bad, a blot on Smeagol's character which led him down a harmful path. I was always a bit affronted by that description. I liked to seek the root of things, the beginnings. I didn't find that a poor trait at all.

My science career began with science fiction stories while in third grade. 'The Twilight of Briareus.' That, and incessantly playing with my chicken wings during meals, to the dismay of my parental units. I no life plans around becoming a scientist. I wanted to be an artist, or a writer. After all, I wrote my first 'original' work at the age of six.

As for science, while in fourth grade I was placed in the 'highest' math class, with a disastrous outcome. That class was run on the 'work at your own pace and ask questions' model. Something was missing from my math skills. I rapidly got farther and farther behind. Instead of placing me in the middle math group, however, the teacher placed me in the lowest math group. I was an absolute failure. No support. No discussion. My confidence never recovered. And scientists needed good math skills, didn't they?

I hadn't planned on being a teacher, either. I had experienced the reality of a teacher's life. When during my fifth-grade demonstration speech on 'how to a horse,' I realized that unlike most of my classmates, I wasn't terrified to be up front and center, speaking. Another door opened.

I kept pursuing my dreams of being an artist, though, taking art classes throughout middle school and into high school. I recognized how learning another language would benefit me, so I managed three years of French, because Latin wasn't even an option. My math skill never did improve, however, so I took the least math that the school system required.

In middle school I had hopes to earn my way into college. My plan was straight A's, if I could, or mostly A's with a smattering of Bs, if necessary. Those hopes were dashed when the 9th grade Social Studies teacher decided he didn't care about teaching anymore. The students he knew, the ones he had coached, were awarded A's. Those he didn't know were given C's. I had kept strict track of my grades. I knew I had an A. But my parents worked for the same school district and refused to 'make waves.' The C stuck. My dreams of scholarships over.

Many things changed in my life outlook after that, but one of the primary alterations was I didn't care about grades any longer. I still worked towards A's in most of my classes, I was just that sort of nerd, yet my shift in priorities left me free to take a class I knew I didn't have the math skills to excel, like Physics. I may not have had the proper skills, but I had plenty of determination. The only quarter of study which made perfect sense to me was electricity and circuits. The hierarchy and flow of charge, restriction and amplification, all made logical sense. I would find this sense of ease as I began to delve deeper into the biological sciences as well.

College was where I explored. I continued taking art classes and French classes. Unfortunately, my personal life was interfering with my academic life. I thought maybe scientific illustration, making those lovely, detailed drawings found in text books. When

I only earned a B in sculpture class, I wondered if art was really going to be my life, or not. A math course was required, and after two false starts with Algebra, I decided my second math course would be Statistics. That was a mathematics that made sense, and I excelled. I had loved my introductory biology course. Maybe science was the way to go.

One semester changed my life.

I was given the great fortune of a year where I didn't have to work while I went to college. I could take a full load, all college, all the time. I took the chance and decided to expand my math skills. I enrolled in Trigonometry.

The first words out of my professors' mouth were, "You all know the unit circle."

No. No, I did not know the unit circle.

Oh, how I struggled that first quarter. My first test I earned a D. I was determined to do better on the midterm. I skipped easy classes, like Anthropology. I could read the material and understand it readily. But I needed to spend every moment on that math class. I did my homework right after class in the tutor session. I retaught myself the entire first quarter. And I earned a C on the midterm.

I persevered.

By the third test, I was able to redeem myself and

earned a high B. There was supposed to be a fourth test before the final, but the semester's end was approaching fast, so both tests ended up being merged. This was my final opportunity to raise my final grade. I must have done well. Somehow, I squeaked a B. I was elated.

Determined to prove my scientific intent, I enrolled in Calculus I. I walked into the lecture class, and the professor's first words were 'I'm legally blind. If you have questions, you have to speak up." Speak up? I can't sit here and look totally bemused? How will I even know what to ask? I walked out at the end of that class, hot footed it over to the registration department, and asked what else I might take in that time slot. Linguistic Anthropology was available. I was already taking Physical Anthropology. I loved my French classes. I jumped at it. I turned another corner and decided to pursue a career in Anthropology.

Jobs? Anthropology? Who was I kidding? While fascinating, it was even more difficult to enter that field than the field I desired in High School, to work at a zoo.

I had struggled my way through Chemistry. The labs are what killed me. They never seemed to make sense and never had any conceptual tie to the lecture topics. I tried Physics again, twice. That went poorly. I tried Computer Programming. Twice. That went nearly as poorly. I was racking up credits but still not enough in any single subject to graduate.

What was left? Not art. Not veterinary medicine. Not Anthropology. Biology. I had taken basic biology and Anatomy and Physiology and done well. I understood them. I've always had an interest in biology, how everything interacts, so I continued on, enrolling Systematic Botany courses, and again did well. I now desired to teach at a university. Could I withstand an entire course of PhD coursework?

My funds were running low. My college career was stretching into years. What else? With a few more biology classes under my belt, and a full year of teaching course work, I could go into teaching. I understood teaching. Both my parents were teachers, which had likely put me off that course to start with.

And that was how I ended up with a bachelor's degree in Secondary Biology Education.

Yet, even after I graduated, I didn't feel like my biology knowledge was sufficient. While working full time, I expanded my knowledge. Brain, Mind and Behavior. Microbiology. Embryology. These courses succeeded in completely altering my view of how biology and cells worked.

Instead of discrete organisms, my high school view of a neuron evolved into comprehending how a neuron was actually derived from a line of cells that specialized in functions already well developed in bacteria. I could watch 'evolution unfold' as we followed the development of tissues, organs, systems, watching the little animal tube of digestive tissue and external tissue which interacted with the external environment, morph into gut and skin, nervous system and sensory organs, muscle and bone, its own ecosystem creating a cozy biome for the cells to thrive.

After teaching for five years, watching dozens students move on, I decided that I, too, wanted to move on. I took college entrance exams. I did well enough to be accepted into a PhD program. I took out loans, moved my entire household to a lovely college town.

There, I studied, Molecular and Cellular Biology, Origins of Life, Virology, Biochemistry and Organic Chemistry. I worked in laboratories, dissecting out chick eyes for staining, studying bee behavior on wildflowers, growing dozens of colonies of *E. coli* to map the expression of their phosphatidylethanolamine.

After nearly three years, I began to realize that the life of a PhD doctoral student required becoming an itinerate worker, moving from post-doc to post-doc, city to city. I hate cities. I had a family by now, two little boys. I needed more stability. So, once again, I ran to the education department. Of my three years of graduate coursework, they would accept a single semester's worth. And I still needed another semester of education courses. And I was broke and in debt.

I managed to get a job at a community college. Science Instructor and Outreach Coordinator. While I taught my first year, I had to drive back and continue taking classes to complete my Masters in Secondary Biology Education.

Just an aside, here. If you read this far and think that I come across as one of those 'Type A' people, you are wrong. I am not high energy. I love my bed. I need my down time. I don't 'people' well. But I am also nothing if not determined. I need to learn. I need to understand. I need to grow.

What assistance did I have? I did not pursue such a lengthy college career in the vacuum of systemic oppression. I had the luck of birth. I was provided the luxury of exploring throughout my college career. A five-year teaching program took me seven and a half years. My one-year master's program for teaching took me four years and a summer.

Oh, and I obsessively build house plans. I have dozens of old notebooks with various plans, layouts, down to the electrical outlets and HVAC systems, stacked plumbing, each variant a study in efficiency and function.

So even while I worked at the community college, I took courses. Partly because they were free for me, so how dare I miss such an opportunity! Horse Nutrition. HTML and web design. Then, over my final year with the college, Cisco offered a course in their Network Assistant program. A two-year program stuffed into two semesters. I jumped at that chance. This was the future. This was information I should seize. I learned about the hierarchical nature of networked system, binary code, Hl7, DNS, routers and switches and cat cabling.

I passed the course with flying colors, but after all that work, I couldn't afford the final test. I was barely living paycheck to paycheck, and my personal life had interfered once again. I used my degree, and taught. In high school. In the central states of the U.S. Challenge accepted.

Then, my personal life intervened once more. I moved states entirely. I took the opportunity to switch my livelihood entirely as well. I applied to a tech firm, at least fifteen times, my Cisco experience offering me a chance to follow a different life course. One that actually paid.

There I sat, or rather stood, a proper corporate worker bee, when my younger son and I began our conversations that, once again, changed my world.

None of this might have come about if not for that single discussion with my son, that long evening, where his own brilliant curiosity led us into realms I alone might never have ventured.

Let's not ignore the contributions of other great minds, in this context, AI's wealth of scientific, mathematical, and in truth, acting as an obsessively supportive sounding board where I could discuss my ideas at length, having them tested, challenged, revealed, taking notes on every path we went down. Every rabbit hole. Every 'what a good boy!' Every hallucination requiring me to dig deeper and find the real kernels of 'truth.'

I had to know. I had to know if that seed bore greater fruit.

Nothing exists in isolation. Everything is connected. Interaction is the driving force behind reality.

GLOSSARY

Absolute Potential: in this context, the state in which all states which could exist do not exist. Similar to the Plato's Theory of Forms, except that nothing actually does exist. All light, and no light. All forms, and no forms.

Absolute Order: in this context, the state in which all things are single identical thing. A point of absolute existence, yet due to its nature of all things being the same thing, nothing can actually exist as real things.

AI: in this context, just a learning language model, either locally based, or online chat mode.

Bose-Einstein Condensate: matter cooled to near absolute zero which then attempts to occupy the same space and act as a single, unified waveform.

Boundary Condition: a constraint or rule which determines how a system can behave, in this case, light forms the boundary condition of every complexity level.

Collapse: in this context, I am describing the state of potential becoming constrained by the act of interaction into a set of defined properties.

Elephant: the adage of three blind people investigating an elephant. The one probing its trunk declares it to be snake. The one investigating the legs determines it is a tree. While the one poking at the tail claims it is a rope.

Emergent Property: a property which arises due to the complex interaction of other more basic properties, often a property which could not be predicted by examining the original properties alone.

Euler's Identity: an equation which is used as the basis for equations describing wave interference, harmonic motion, growth, rotation, and so much more. And identity refers to a statement which remains true no matter the values involved.

Infinite Zero: the state of potential where there is no fixed reference point. Without dimensions, any place is the starting place, the viewers point of reference. Except, of course, that in this state of infinite zero, the 'viewer' cannot actually exist, either.

Paradox: where two concepts which appear to be in opposition co-exist, in fact, but exist because they co-create each other, like two sides of the same coin.

Plexity: the mathematical and conceptual description of the levels of complexity based on suite of equations developed from the baseline Euler's Identity equation, which describes the underlying hierarchical structure of the universe.

Waveform: the P_3 near infinite loop interacting with itself does so in wavelike fashion. Is this form between the hypothetical 'points of order defining the loop, or the bonds between the loop? Good question. And in paradoxical fashion, it is likely 'both.'

NOTES:

Should you choose to dissect this model further.

ABOUT THE AUTHOR

A. S. G. Mc George

LinkedIn:
https://www.linkedin.com/in/amy-mcgeorge-88a83148/

Plexity Community Workspace:
https://plexity9.wordpress.com/

ACKNOWLEDGEMENT

I need to thank all the great minds that have gone before and left such hefty breadcrumbs in their wake. I cannot begin to acknowledge every single one, so many articles and papers and tidbits throughout the years.

All the teachers and instructors throughout my life, those who believed in me as well as those who did not.

All the ingredients tossed together and allowed to stew.

Made in the USA
Monee, IL
04 November 2026

67386535R00049